HERBERT S. ZIM

THE UNIVERSE

NEWLY REVISED EDITION

illustrated with drawings
by Gustav Schrotter and René Martin
and photographs

WILLIAM MORROW AND COMPANY
New York 1973

Copyright © 1961, 1973 by Herbert S. Zim
All rights reserved. No part of this book may be reproduced or utilized in any form or by any means, electronic or mechanical, including photocopying, recording or by any information storage and retrieval system, without permission in writing from the Publisher. Inquiries should be addressed to William Morrow and Company, Inc., 105 Madison Ave., New York, N.Y. 10016.
Printed in the United States of America.
Library of Congress Catalog Card Number 72-14200
ISBN 0-688-30096-0 (lib. bdg.), ISBN 0-688-25096-3 (pbk.)
3 4 5 77 76 75

The author is especially grateful to Dr. Stanley P. Wyatt of the University of Illinois, Department of Astronomy, who graciously read the manuscript and offered his advice and comments. James R. Skelly helped materially on the revision.

The photographs come from California Institute of Technology, University of California, Harvard College Observatory, and Yerkes Observatory.

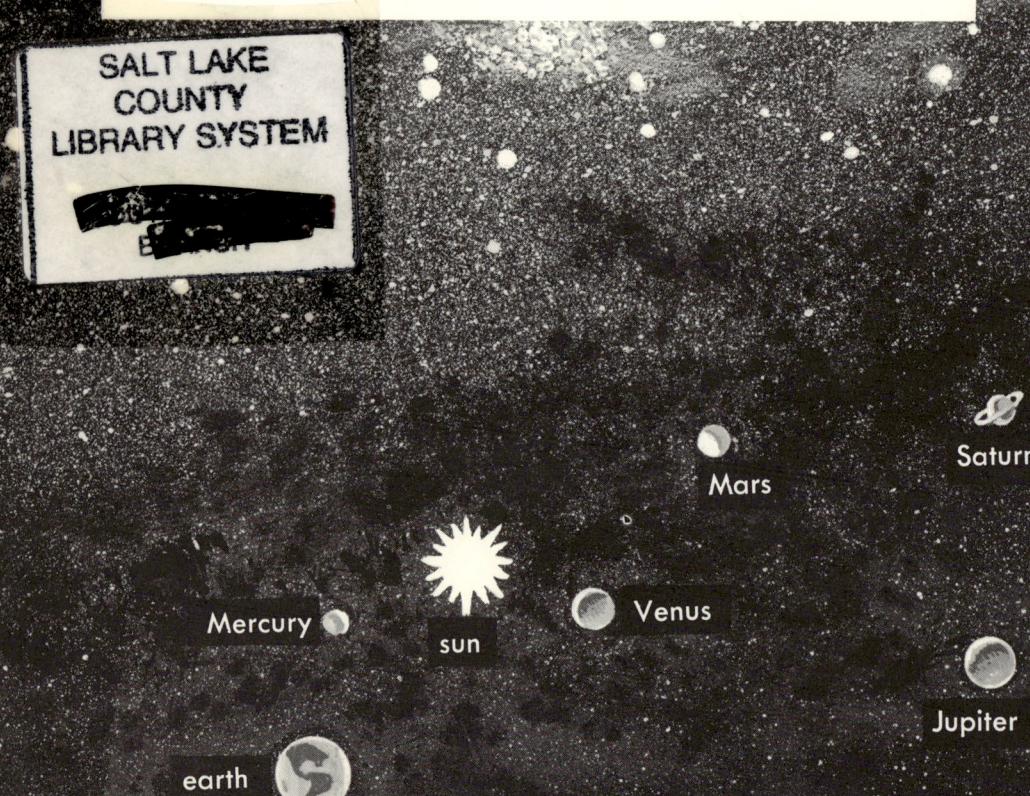

galaxies in the constellation Leo, the Lion

People have always been curious about big things — the tallest mountain, the longest river, the largest dinosaur, the deepest ocean, the greatest diamond. But whatever you name, one thing always has been and always will be the biggest of them all. That is the universe. By its very definition the universe includes everything. So anything you name can only be part of the universe.

Big though it is, the universe may be getting even bigger. Some scientists believe it is expanding minute by minute. More importantly, our knowledge of the universe seems to be growing even faster. The universe of today is much larger than the universe of 1800. And the universe of 1800 was many times larger than that of a thousand years before. So we live in an ever-changing universe. The more astronomers study it, the more they discover and add to it.

FOUR GREAT TELESCOPES

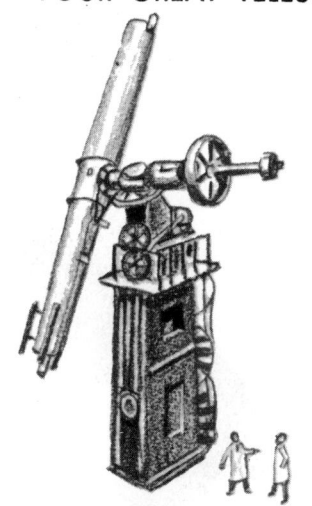

U.S. Naval Observatory 26-inch refractor

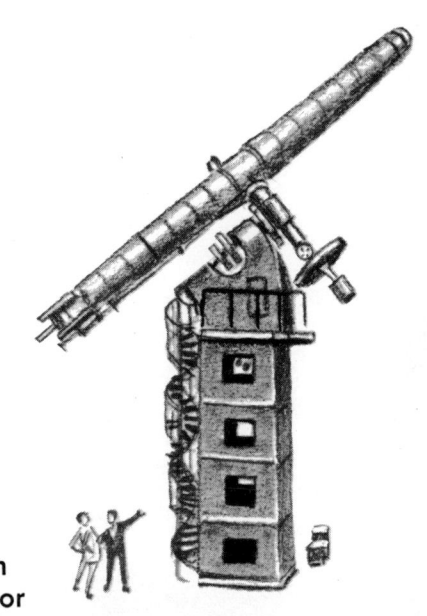

Yerkes 40-inch refractor

The universe we know today is so large that we cannot say in easily understood words or numbers just how big it is. Over 99 percent of the universe is invisible, even to astronomers with telescopes. Now radio telescopes probe farther into space. Telescopes carried well above the earth by space satellites give clearer pictures, and telescopes placed on space stations or on the moon may do even better. Still, many critical questions about the universe are unanswered.

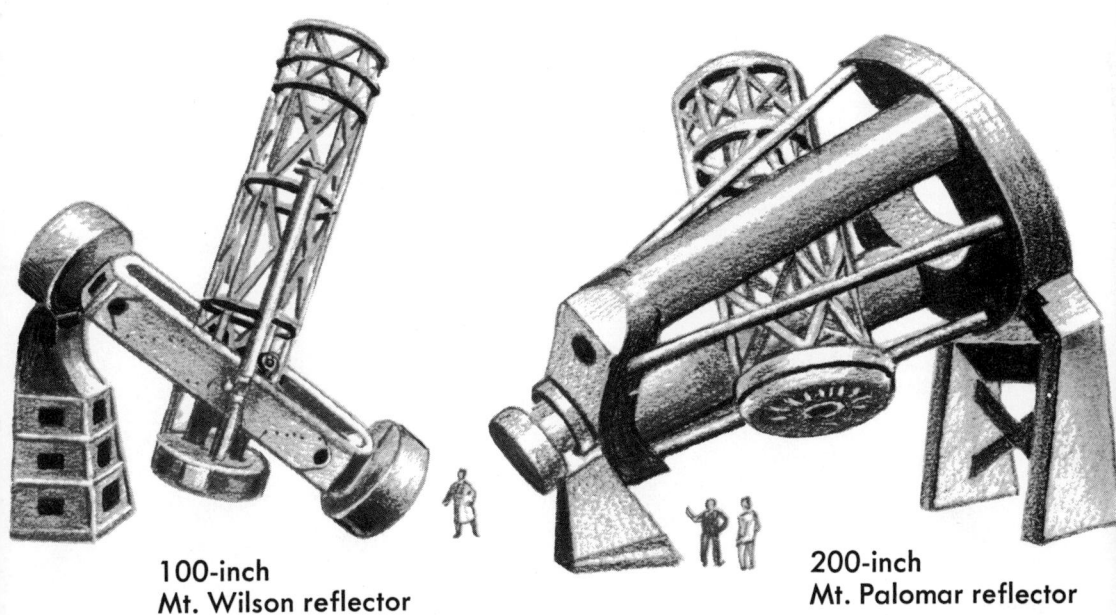

100-inch
Mt. Wilson reflector

200-inch
Mt. Palomar reflector

To find out about the universe, scientists use some of the most important tools ever invented. The first is the ruler, a tool that makes all measurement of distance possible. The second is the clock, which measures time. Measurements of time and distance go hand in hand in the study of the universe. Other important tools are scales and balances, lenses and mirrors, film, and instruments for detecting and measuring light and other radiant energy.

basic tools to measure time, distance radiation, and weight

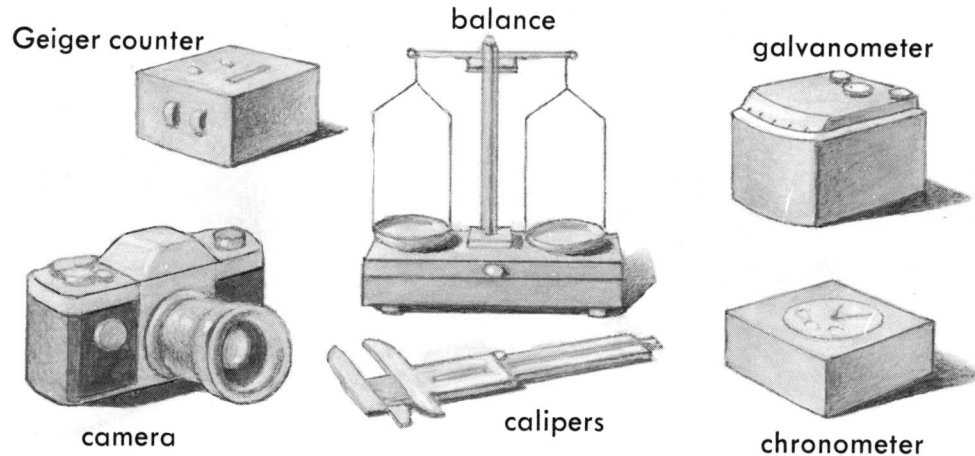

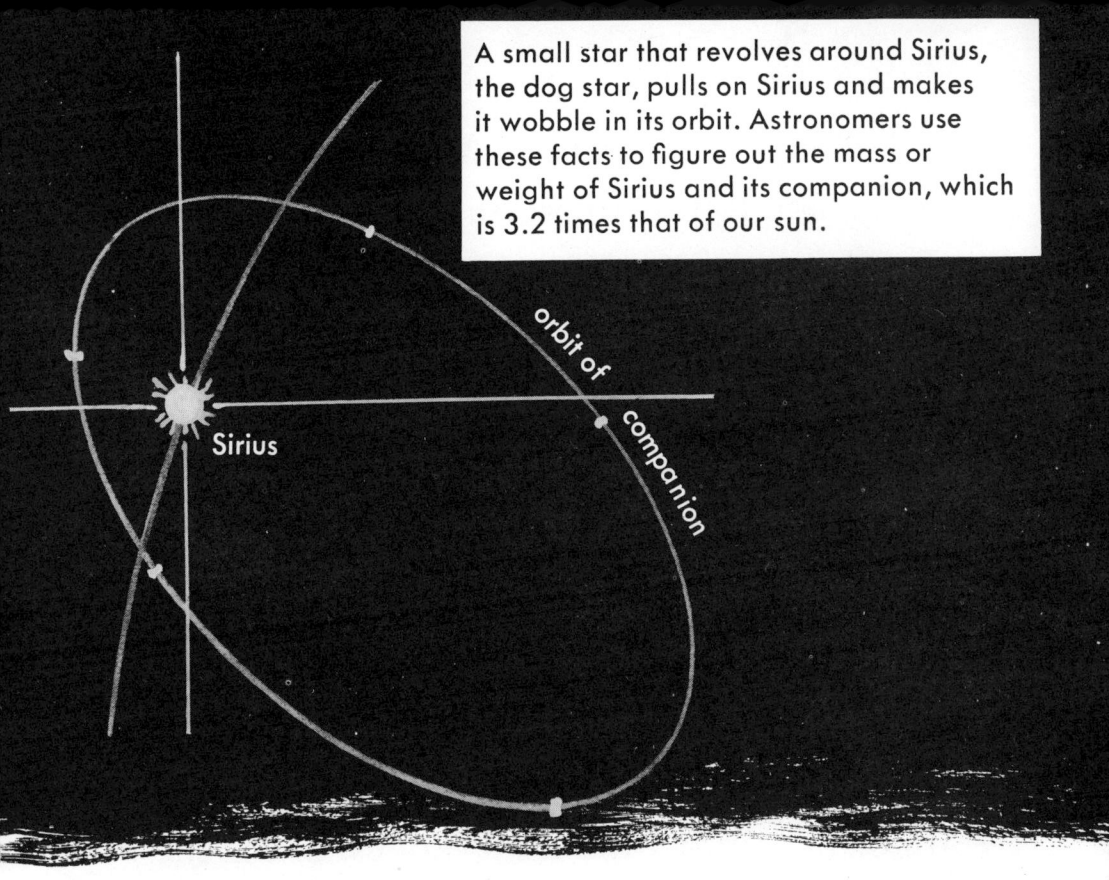

A small star that revolves around Sirius, the dog star, pulls on Sirius and makes it wobble in its orbit. Astronomers use these facts to figure out the mass or weight of Sirius and its companion, which is 3.2 times that of our sun.

Astronomers cannot weigh the universe or any major part of it, but they can compute the weight of distant stars indirectly. Observations with telescopes give clues to the motion, weight, distance, age, and composition of stars and other bodies that are the material of the universe.

For over a million years people have wondered about the universe and have tried to picture what it is like. Ancient people had strange ideas that may seem foolish to you and me. But these first ideas show how long men have been thinking about their universe.

Some American Indians thought that the universe rested on the back of a giant turtle swimming in the sea. The turtle had dived to the bottom and brought up mud, which it formed into hills and valleys. Since these Indians never traveled far, they thought the nearby hills and valleys, rivers and forests made up the whole universe.

Egyptians who lived in the Nile valley 5000 years ago thought that their valley was the universe. Around this rich land were tall mountains, which held up the sky. And when the great barge of the sun-god moved in front of the mountains, it was day. When it turned behind the mountains, it was night.

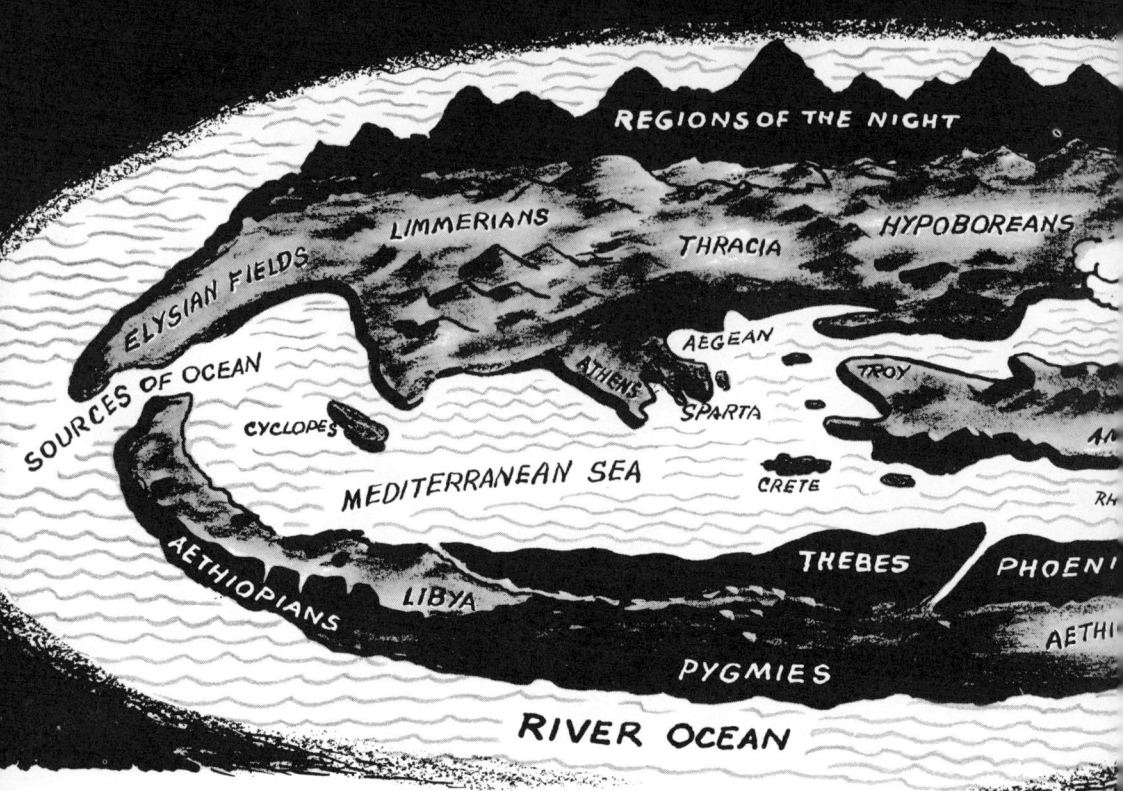

A thousand years later, most Greeks pictured the same kind of universe. At the very center was a great sea called the Mediterranean, which means middle of the earth. The sea flowed out through what we now call the Straits of Gibraltar. Then it formed a great river, which encircled the universe.

Around the Mediterranean Sea were all the lands of the world. The Greeks, who sailed

this inland sea, lived to the north of it. To the south lived Pygmies and other dark-skinned people. The sun-god drove his chariot across the sky, disappearing into the ocean at evening. The stars followed in his path.

About 3000 B.C., these ideas about the universe began to change. A few Greeks, who might be called the first scientists, started to observe the earth and sky with much more care. As they studied they came to see that the universe was not as most people believed. To explain what they saw and measured, these Greek scientists knew their universe had to be much larger.

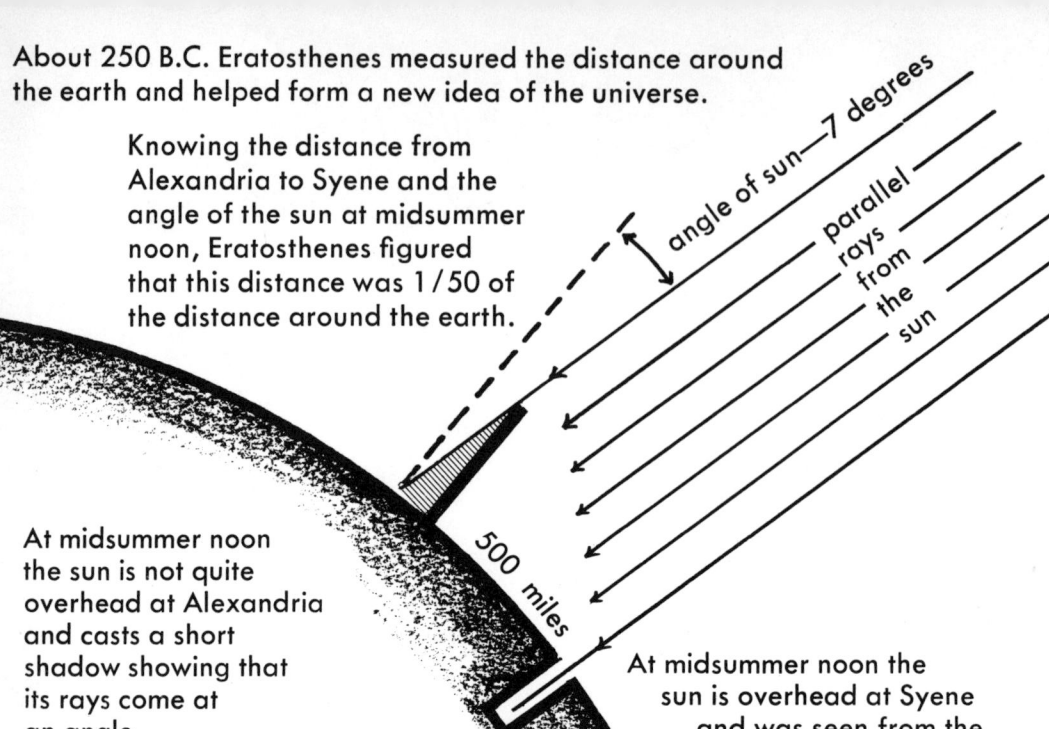

About 250 B.C. Eratosthenes measured the distance around the earth and helped form a new idea of the universe.

Knowing the distance from Alexandria to Syene and the angle of the sun at midsummer noon, Eratosthenes figured that this distance was 1/50 of the distance around the earth.

At midsummer noon the sun is not quite overhead at Alexandria and casts a short shadow showing that its rays come at an angle.

At midsummer noon the sun is overhead at Syene and was seen from the bottom of a well.

These Greeks first saw the earth as a ball. They reasoned that it was at the center of the universe, with the sun, moon, planets, and stars going around it. Hipparchus made a catalog of stars. He measured the distance to the moon and studied the motion of the sun and planets. From these facts came a system that made events such as an eclipse of the sun predictable.

Later, in the second century A.D., a great Greek astronomer named Ptolemy gathered together many ancient observations and added his own. Skilled in the use of geometry, Ptolemy developed the system that later carried his name. He set up a model of the heavens and tried to account for all its movements. This task was difficult, for some planets seemed to reverse their direction for a short distance before going forward again.

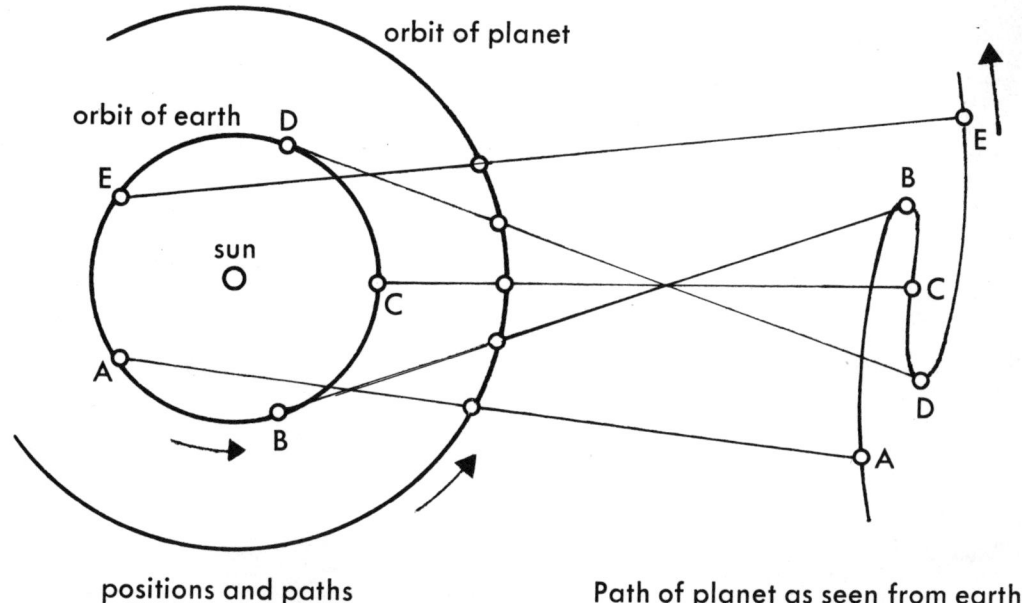

positions and paths
of earth and planet

Path of planet as seen from earth
seems to reverse direction.

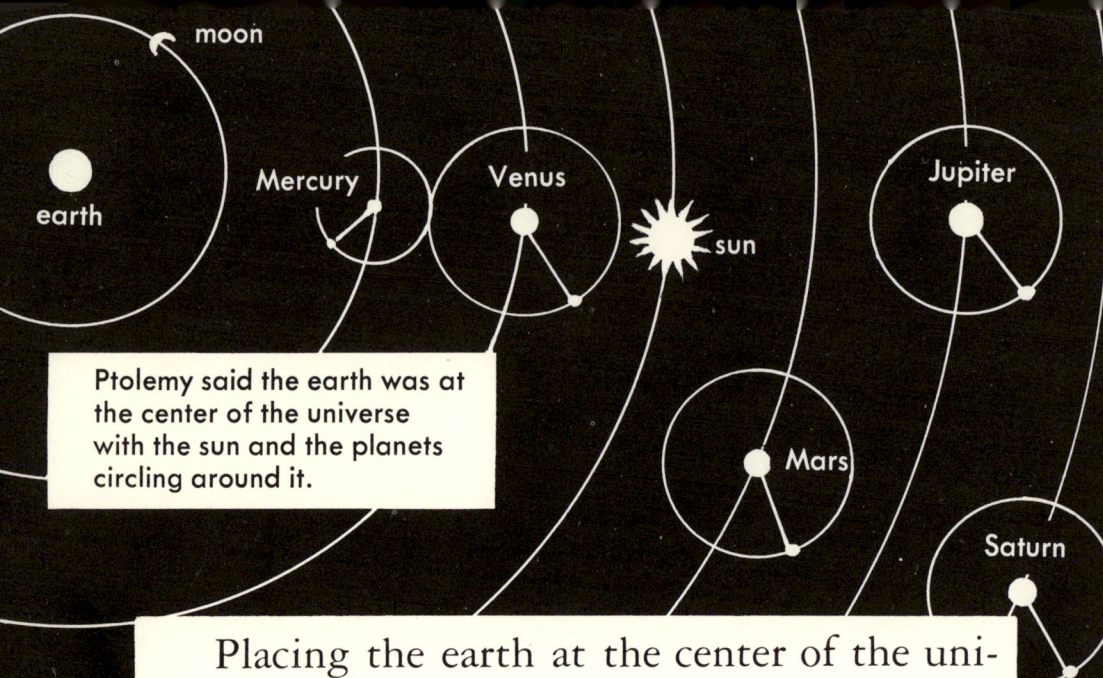

Ptolemy said the earth was at the center of the universe with the sun and the planets circling around it.

Placing the earth at the center of the universe, Ptolemy devised complicated orbits, or paths, around it in which the sun, moon, and planets moved. In the first orbit came the moon, followed by orbits of Mercury, Venus, the sun, Mars, Jupiter, and Saturn. Finally, around all was the daily-turning sphere of stars. Most important of the stars were 12 groups or constellations of about equal size. These formed the zodiac, a belt in which the sun and planets seemed to move month after month.

Such a picture of the universe matched what people could see with their own eyes. It explained the movements of the stars, the sun, and planets much better than before. So everyone who knew about them accepted Ptolemy's views. Because they were in harmony with the well-known writings of Aristotle, Ptolemy's ideas were taught in all the schools. For the next 1400 years his model of the universe was accepted by the most learned men and nearly everyone else.

A recently discovered Greek mechanical computer was made over 2000 years ago. It used geared wheels to predict the movement of the moon and planets.

Brahe made accurate observations of stars and planets, even though he had no telescope.

But during these years some men looked further at the heavens, thought for themselves, and asked questions. Tycho Brahe, a great Danish astronomer, observed stars and planets night after night in the late 1500's and kept detailed records of their positions and movements. From these notes, Johannes Kepler developed three laws describing the motions of planets still valid today. His work was part of a great spurt of scientific research that began in the early 1600's.

Dutch experimenters found that curved pieces of glass made distant objects look closer. Galileo Galilei, a professor at Padua, Italy, improved on this idea and made the first practical telescope. On January 7, 1610, he turned his small telescope on the moon and saw the craters distinctly. Galileo looked at Jupiter and discovered four moons traveling around the giant planet. With the telescope, one discovery quickly led to another. Clearly Ptolemy's old model of the universe could not explain the newer discoveries made with the telescope.

Galileo made his first telescope in 1609.

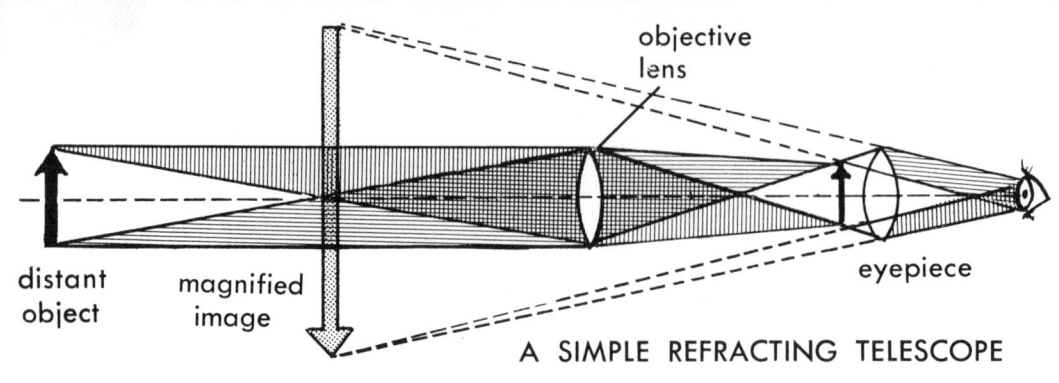

A SIMPLE REFRACTING TELESCOPE

But by now the ideas of Ptolemy and Aristotle were so widely accepted that to believe anything but what these masters had taught was almost criminal. In those days, people were punished for their beliefs if they differed from the official teaching—something much less common today. Copernicus, the famed Polish astronomer who thought the sun—not the earth—was the center of the system of planets, did not dare publish his book until the year of his death, more than thirty years after it was written. Giordano Bruno was burned at the stake in 1600, because, among other things, he denied that the earth was the center of the universe.

The new discoveries started astronomers thinking about ideas they had long taken for granted. By the late 1600's the books of Kepler and Copernicus were studied openly. Astronomers compared what they read with what they saw through their telescopes. New ideas about the universe began to spread.

Other scientists helped by improving the telescope. In England, Sir Isaac Newton worked out the laws of motion and of gravitation. They accounted for the observed movements of heavenly bodies. In 1781, the planet Uranus was found by accident. In 1843 and 1846, two mathematicians each

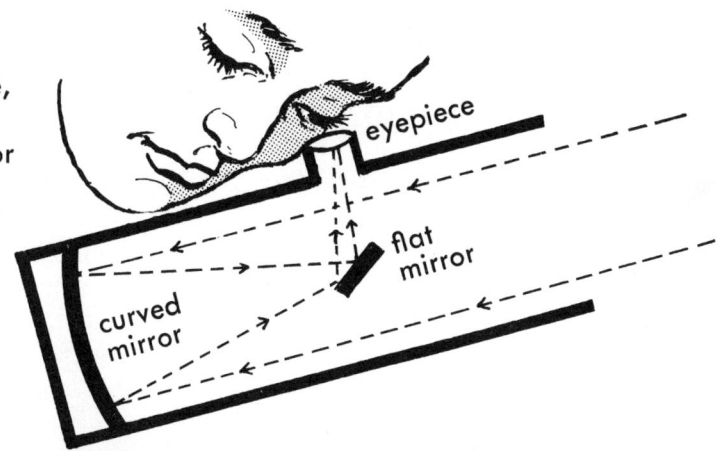

Newton's telescope, invented in 1670, used a curved mirror instead of a lens.

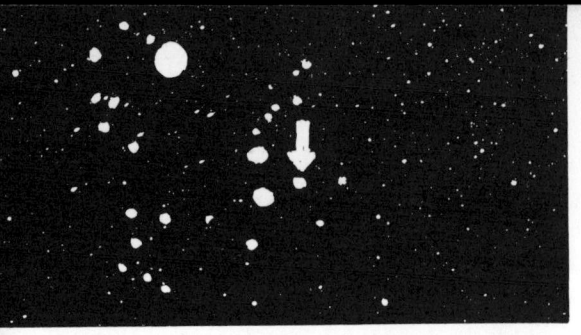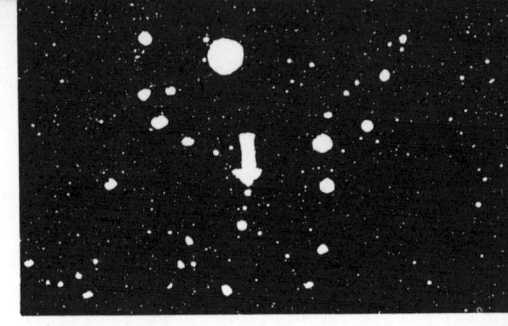

The planet Pluto was discovered by the careful study of detailed photographs.

predicted the discovery of the planet Neptune and told astronomers exactly where to look. After a twenty-year search based on detailed studies of the movements of Neptune and Uranus, the discovery of Pluto came in 1930.

As astronomers peered into the sky with bigger telescopes they discovered many small planets (asteroids) between Mars and Jupiter. They found more moons traveling around the major planets. They studied comets and meteors — minor members of the solar system. Finally, they turned their telescopes on the stars. Even the brightest ones, however, only looked like points of light.

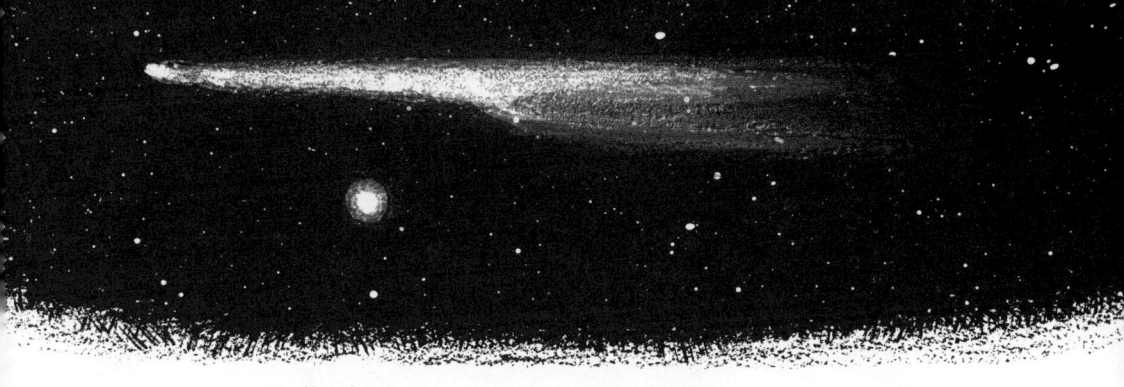

Halley's Comet near Venus in 1910

Thus, over a period of three or four thousand years, our ideas of the universe grew and grew. Through discovery and exploration people came to think of the universe as an orderly system centering on the sun and its planets. By 1800 the main pattern of the solar system was well known, though any astronomer will tell you there is still much to learn about comets, meteors, satellites, and even the other planets.

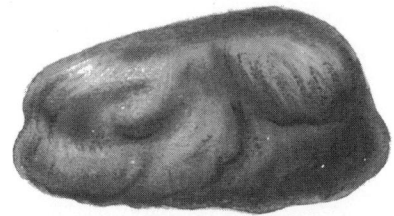

stony meteorite

iron meteorite

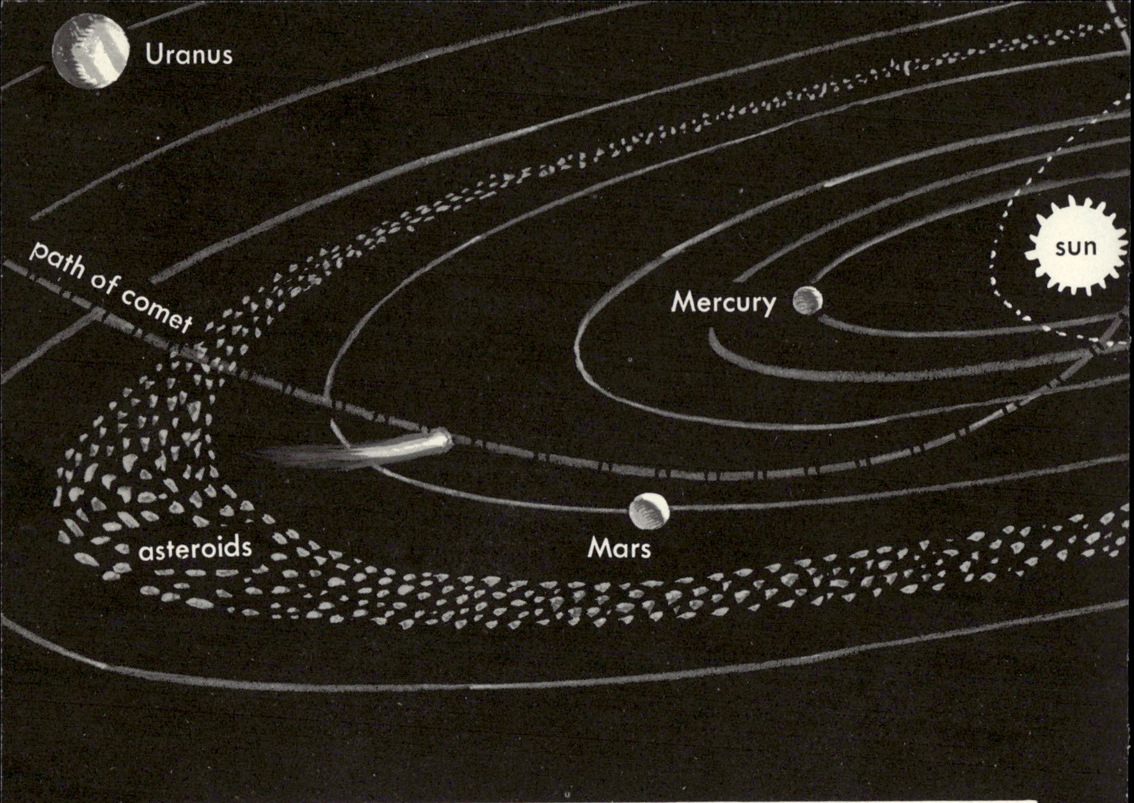

However, even before Copernicus and Galileo, some people felt that there was much more to the universe than the part we call the solar system. They guessed that the stars might be suns like our own, but much farther away. Newton's ideas on motion and gravitation helped turn the attention of astronomers from the planets to the stars.

THE SOLAR SYSTEM

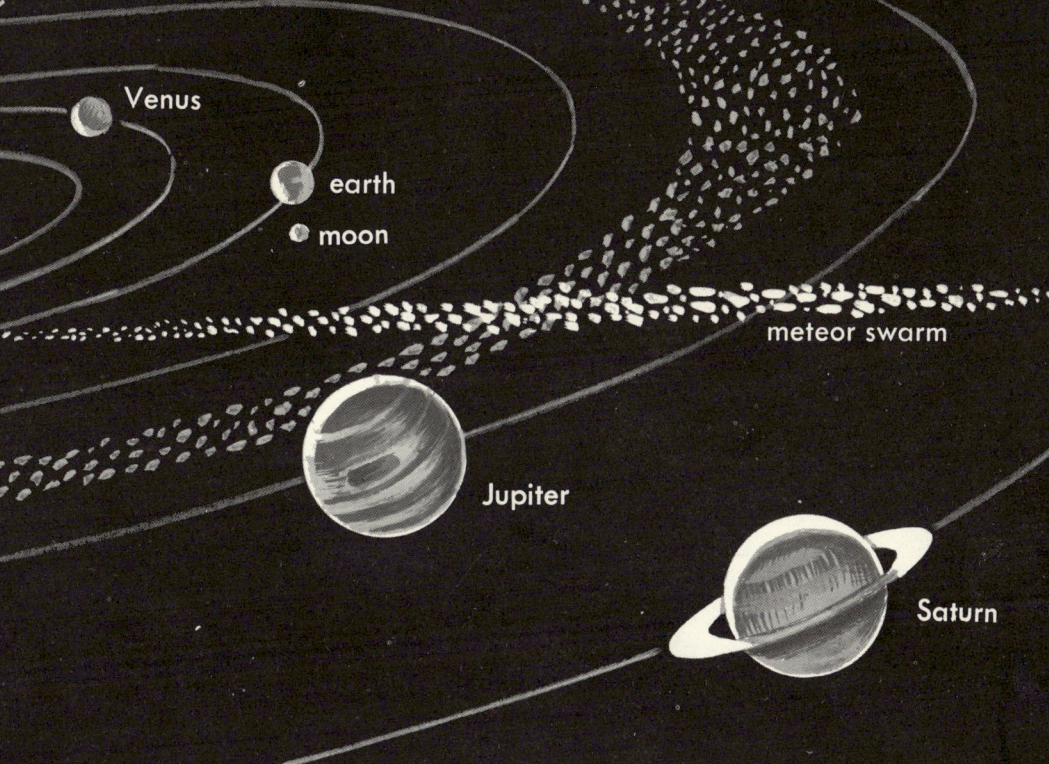

The stars, of course, had been studied for centuries. Those that they could see, the Greeks arranged in patterns, or constellations, to represent their gods, goddesses, and heroes. The Arabs were less interested in constellations. They concentrated on the stars themselves, giving the brighter ones Arabic names that are still used.

SOME OF THE BRIGHTEST STARS seen in the northern hemisphere			
Name	Constellation	Brightness	Color
Sirius	Big Dog	−1.4	white
Vega	Lyre	0.0	white
Arcturus	Herdsman	0.1	orange
Capella	Charioteer	0.1	yellow
Rigel	Orion	0.1	blue-white
Procyon	Little Dog	0.4	yellow-white
Betelgeuse	Orion	0.4	red
Altair	Eagle	0.8	white
Aldebaran	Bull	0.9	orange
Spica	Virgin	0.9	white
Antares	Scorpion	0.9	red

(brightness—the lower the number the brighter the star)

Later astronomers studied and restudied the stars with telescopes. Some drew maps of the fainter stars. Others discovered clouds of glowing gas, which they called nebulae. Some stars were discovered to be twins. Some faded and grew brighter on a regular schedule. On rare occasions a star would increase rapidly in brightness, and then fade out and disappear.

Great Nebula in the constellation Orion

Yet with all the interest in astronomy, no one was able to measure the distance of stars. Finally, from 1837 to 1839 three astronomers succeeded. The idea they used was simple. Hold a pencil up a foot or so in front of you and look at it, first with one eye and then the other. The pencil seems to move across the background, because your eyes are about three inches apart and so give you two slightly different points of view. If a friend holds the pencil up across the room, you will find that the shift is much less. Farther away the shift, or parallax, can scarcely be seen.

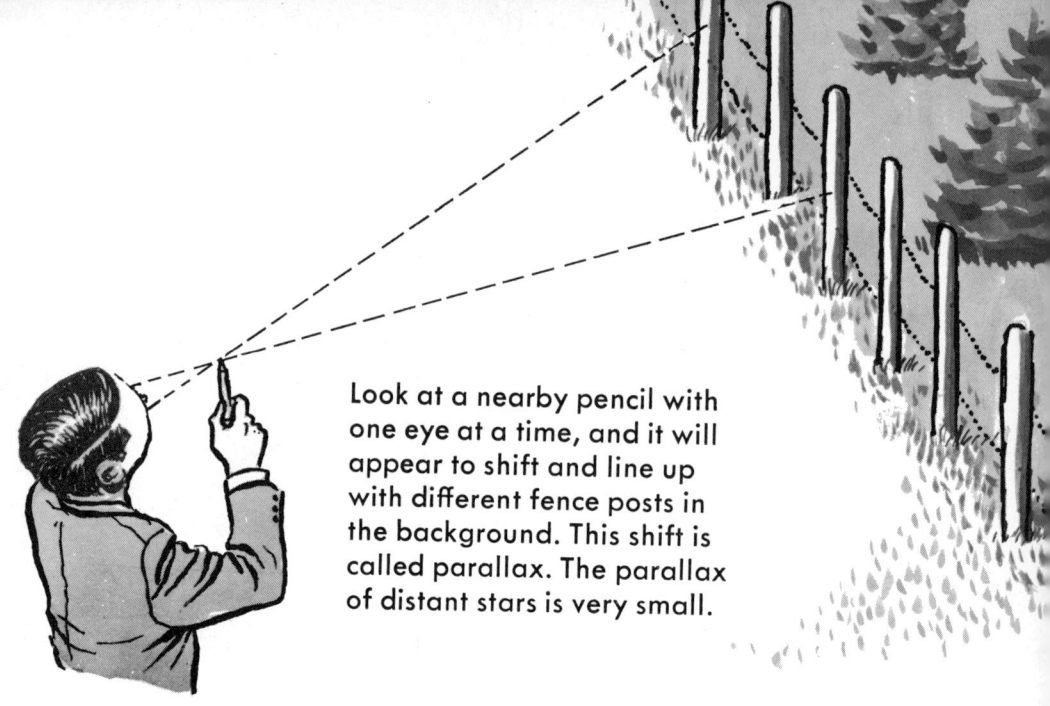

Look at a nearby pencil with one eye at a time, and it will appear to shift and line up with different fence posts in the background. This shift is called parallax. The parallax of distant stars is very small.

By measuring the shift against the background, you could, if you wished, figure out how far the pencil was from your eyes.

Bessel, one of the astronomers, carefully drew maps of stars in the constellation Cygnus the Swan. He did one in the spring and another in the fall. While these six months passed, the earth had moved halfway around the sun and was 186 million miles from its first position. Thus Bessel could look

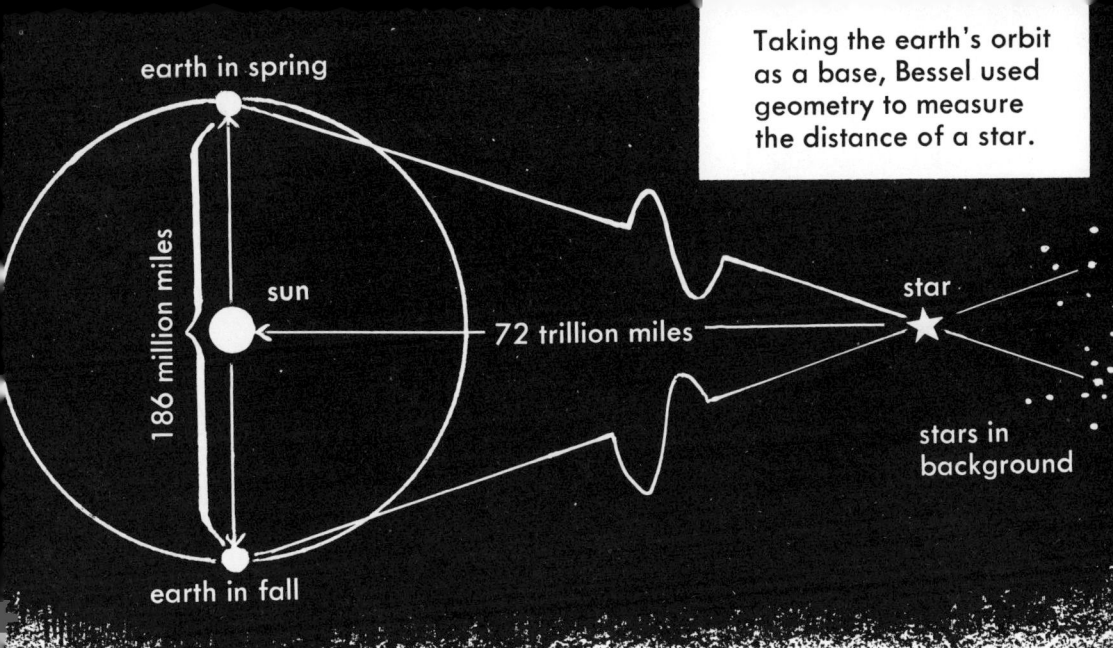

at his stars from two points of view — quite a distance apart. He searched for a shift of one star's position against the background of the others — even a slight one — so that he could compute the distance of the star.

At last Bessel was successful. The first star he measured turned out to be 72 trillion miles away. This figure was hard to believe. Other astronomers searched and found closer stars, but none closer than 26 trillion miles.

This simple method of measuring the distance of stars set the boundaries of the universe out about 100,000 times farther than before. As the universe began to unfold, more attention was given to distant stars. The solar system seemed less important.

But even before the distance of stars was first measured, astronomers were curious about the great band of stars across the sky that we call the Milky Way. Some guessed that the Milky Way marked the boundaries of the universe, but they were not sure. In 1784, Sir William Herschel and his sister Carolyn, both famous astronomers, began a long and detailed study of the Milky Way.

a section of
the Milky Way

If stars were spread evenly through a universe shaped like a ball, you could see the same number no matter which direction you looked.

To understand what the Herschels found, imagine yourself in the center of a gigantic ball containing millions of stars all about the same distance apart. No matter which way you look you see about an equal number of stars, because you are in the center and the stars are spread out evenly. If the ball was stretched until it became football-shaped, the appearance of the sky would change, even though the number of stars did not. Now, looking through the long axis of the football, you would see more stars than if you looked up the short way.

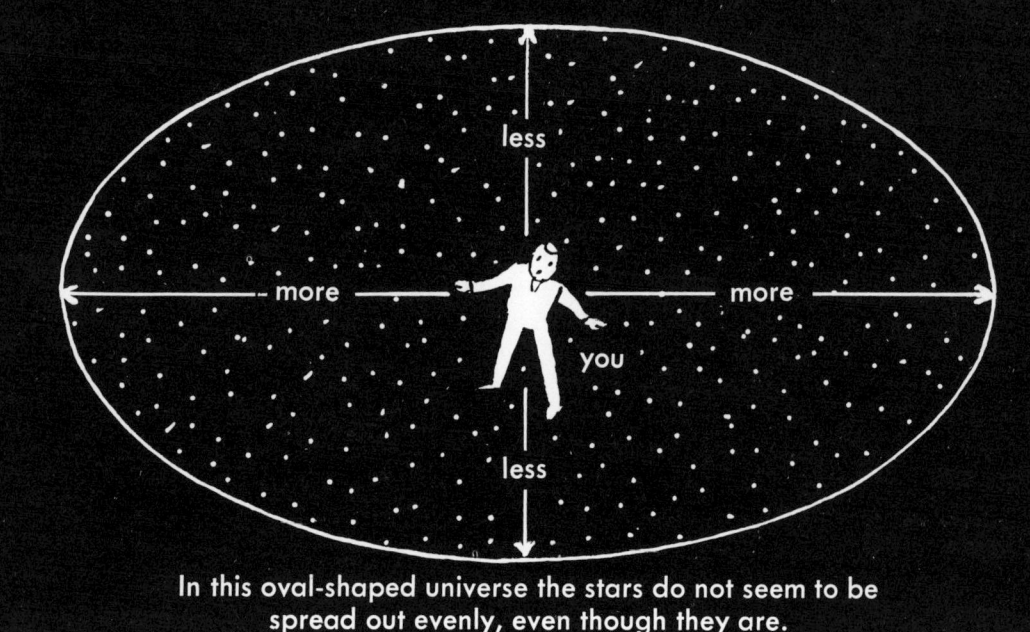

In this oval-shaped universe the stars do not seem to be spread out evenly, even though they are.

If your ball of stars was flattened out until it became as thin as a wheel, or disc, you would see more star-studded space if you looked toward the edges of the wheel. You would see fewer stars if you looked up at right angles to the wheel. The great family of stars that includes our sun is mainly in the shape of a disc. As you look up across the disc you see few stars. As you look toward the edge of the disc you see many. This view is what we call the Milky Way.

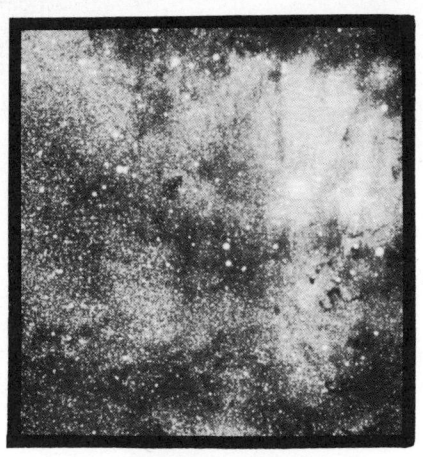

photograph taken through the Milky Way

photograph taken at right angles to the Milky Way

Since 1750 astronomers had guessed that the sun belonged to this disc-shaped system now called a galaxy. William Herschel, his sister, and others mapped and measured hundreds of its brighter stars. These data supported the idea of a system of stars, but a century passed before the full story was developed. Improved telescopes, cameras, and other equipment helped, because the critical task was still to measure the distances of the faintest stars.

The horizons of astronomers grew so much in the 1800's that new ways of calculating great distances were needed. Measuring the distance to the stars in miles is as difficult as measuring the distance across Africa in inches. Instead of a mile, astronomers picked the average distance of the earth to the sun — 93 million miles — as a unit of measurement. They called it the astronomical unit (AU). Saturn is 886 million miles from the sun. On this scale the distance is almost 10 astronomical units.

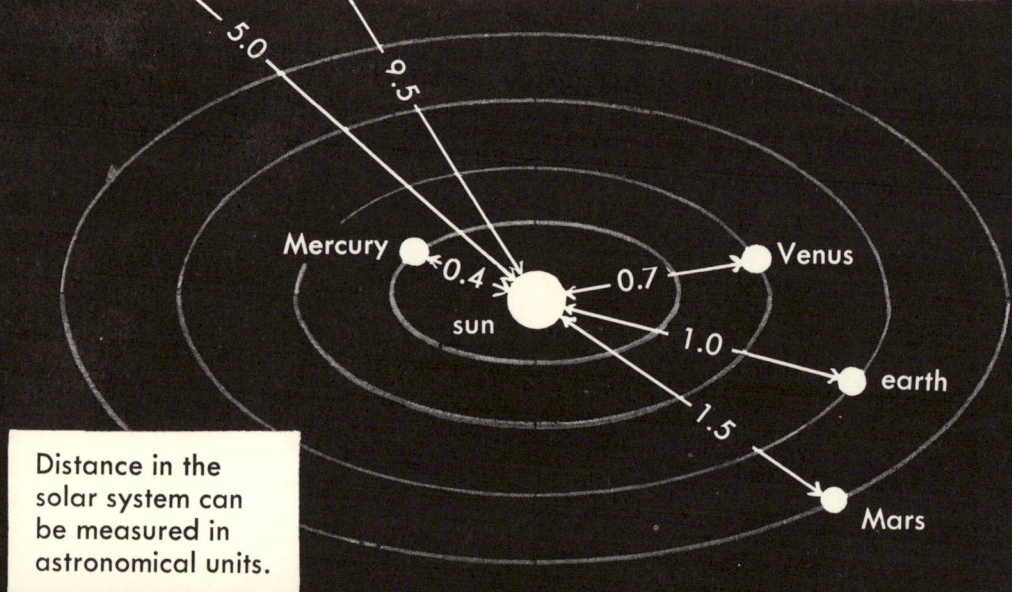

Distance in the solar system can be measured in astronomical units.

When this unit proved to be much too small, astronomers turned to light and time for a solution. You and I measure distance by the foot, the meter, or the kilometer. Ordinarily this system works very well. But you might use a unit of time instead of a unit of distance. If two cities 60 miles apart are connected by a road on which you can drive safely at 60 miles an hour, you might say the cities are an hour apart. Sometimes we say a place is five minutes down the road. Airlines often use time for distance. Miami, Florida, and New York City are about 2 hours apart by passenger jets, which fly 600 miles an hour.

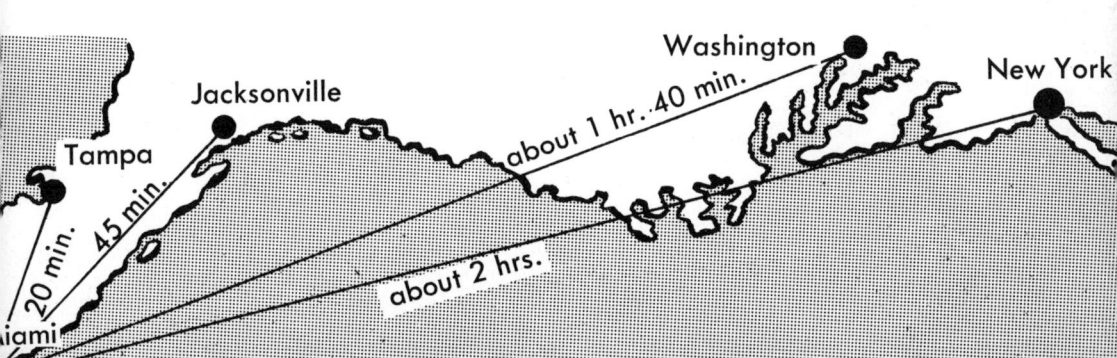

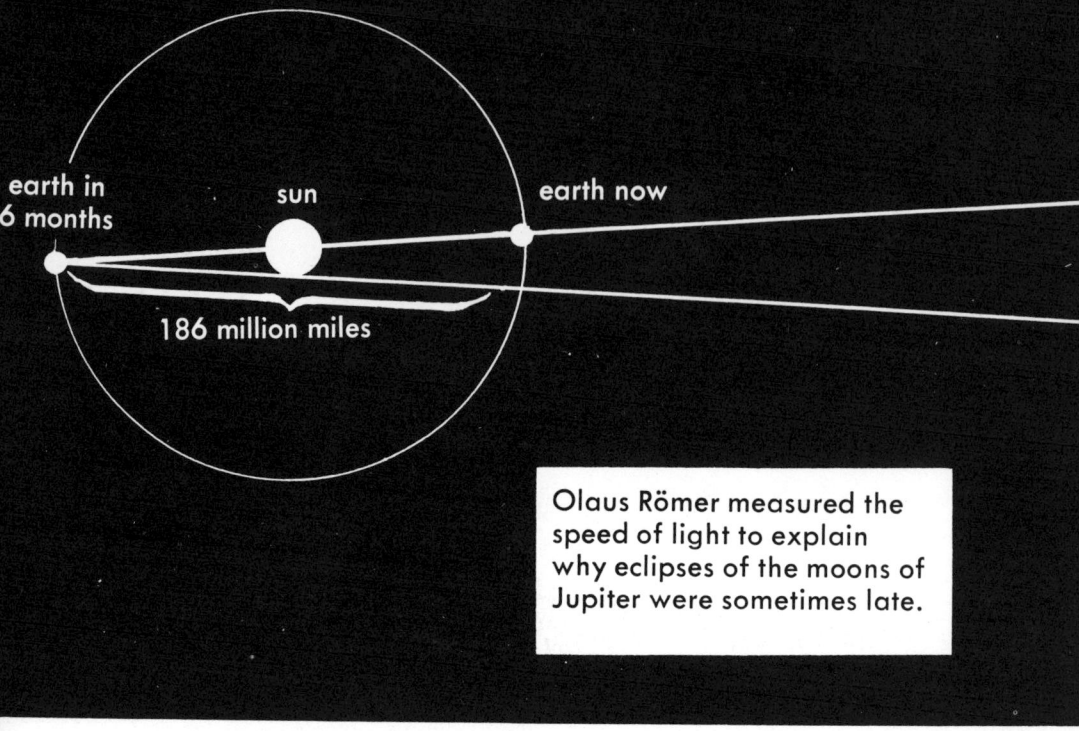

Olaus Römer measured the speed of light to explain why eclipses of the moons of Jupiter were sometimes late.

For astronomical distances, the speed of light, not the speed of planes, is used. The speed of light is good for a unit of distance. It is the fastest speed known, and it does not change in outer space. Olaus Römer measured it at 186,000 miles *a second* in 1675. He was very accurate for his time. Today scientists measure it with much greater accuracy at 186,282.3960 miles a second.

At this distance from the earth the eclipse of a moon of Jupiter takes place every 42½ hours.

Jupiter now

Six months later light from Jupiter has to travel 186 million miles farther. It takes 1000 seconds (16 minutes) to do so.

At this distance an eclipse of the moon comes 16 minutes later than expected.

Jupiter in 6 months

Light seconds, minutes, or hours will express distances in the solar system, but the unit that astronomers use when studying the stars is the light-year (LY). This distance is tremendous. Each year has some 31,560,000 seconds, and in each second light travels 186,000 miles. Thus, a light-year is equal to a distance of nearly six trillion miles or over 63,000 astronomical units.

Can you imagine the distance of one light-year? Once around the earth is 25,000 miles. This length comes to 1/7 of a light second. From the earth to the moon — 250,000 miles — is not quite 1 1/2 light seconds. The 93

DISTANCES FROM THE SUN TO PLANETS AND STARS
in light-minutes, hours, and years

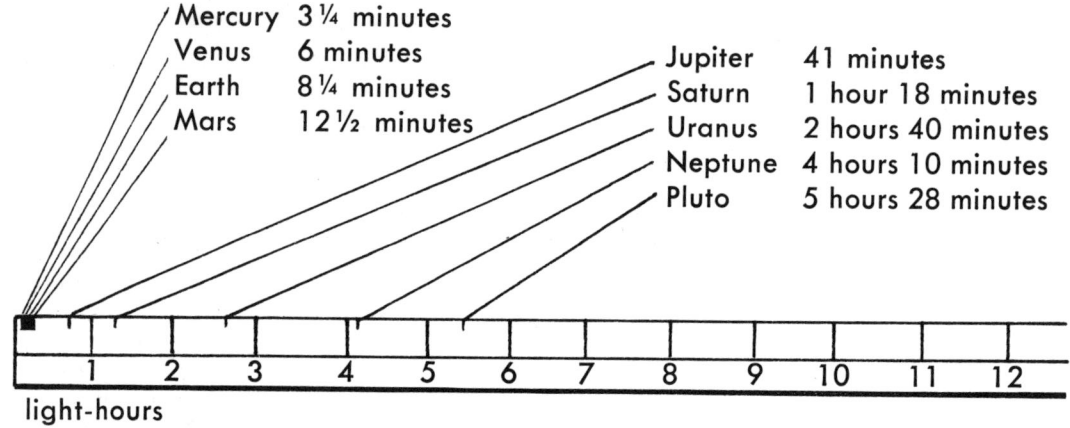

million miles from the earth to the sun, 1 astronomical unit, comes to a bit over 8 light minutes. The diameter of the whole solar system is not quite one half of a light day, or about 1/1000 of a light-year. Yet the nearest star is so far that its light takes 4.3 years to reach us. Sirius, the brightest visible star, is 8 light-years away, 50 trillion miles.

More widely used in astronomy is a measure that depends on the shift in a star's position when observed from different points of the earth's orbit. The farther the star, the smaller the shift, or parallax. When the shift is 1 second (less than 1 millionth of a circle) the distance of the star is 1 parsec, about 3.3 light-years.

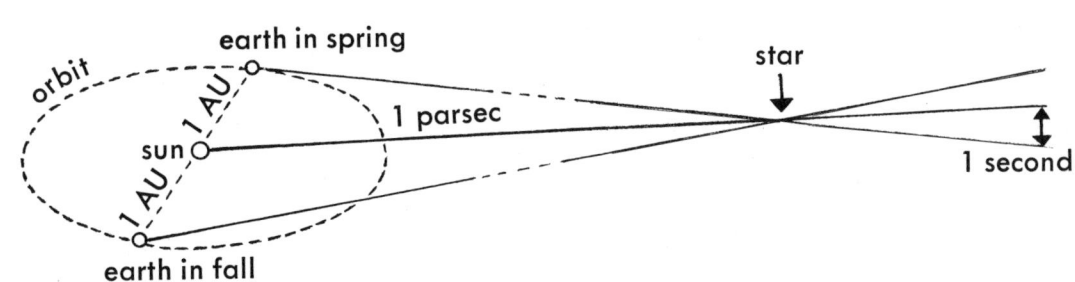

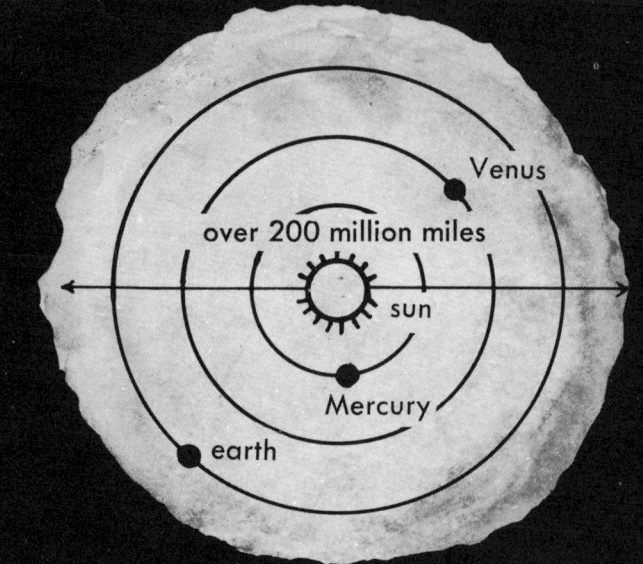

Some red giant stars are 200 million miles in diameter.

Our galaxy is made up of some hundred billion stars. Many are larger and hotter than our sun, which is average in size. Stars range from small, dense white dwarfs to the largest red giants, which are so big that one would more than fill the earth's orbit.

Overall our galaxy seems to be a vast ball some 100,000 light-years across. Through much of the galaxy stars are few and far between, but many more are found across the center in a flat spiral disc that forms the best-known part of the galaxy.

This thin disc is also some 100,000 light-years across. At its inner core, or nucleus, it bulges out to a thickness of some 5000 light-years. The galaxy disc with its spiral arms is a star-studded pinwheel that revolves at a rate of about 200 miles a second. At the very center and at the tip of the arms, the speed is less.

In the galaxy are gases and dust and many, many stars. In addition to the kinds of stars that occur throughout the galaxy, the spiral arms have a special population of bright supergiant stars, open star clusters, and other types. In one arm, also, about 30,000 light-years out, is our own small sun.

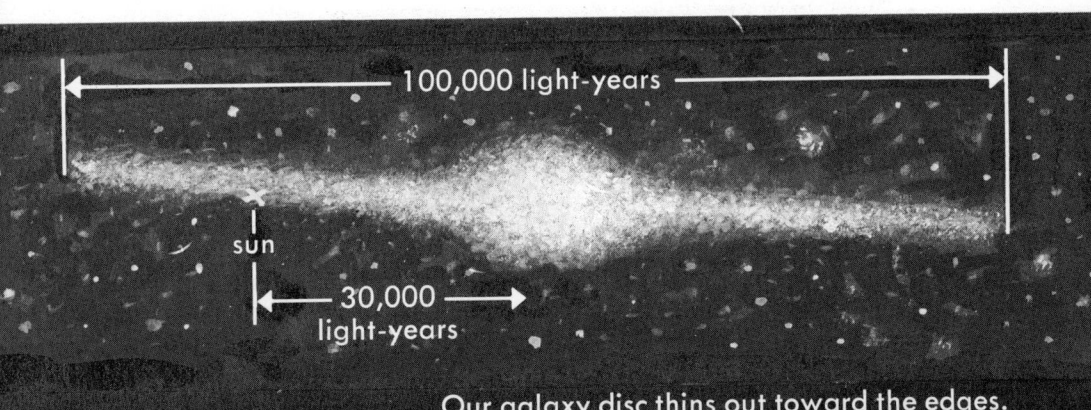

Our galaxy disc thins out toward the edges.

Our galaxy is a flattened spiral containing about 100 billion stars. The position of the solar system is marked X.

The sun has its own orbit, or path, in the galaxy. While our earth takes one year to orbit the sun, the sun takes two hundred million years to orbit the galaxy.

In the disc of the galaxy are at least a thousand open clusters of stars. Probably many more are hidden by clouds of dust and gases. An open cluster may contain 20 to 30 stars or perhaps 10 times that number. Most clusters include very bright stars, like those in the Pleiades, the best-known open cluster.

The Pleiades are an open cluster of stars within our galaxy.

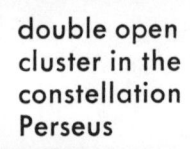

double open cluster in the constellation Perseus

Also in our galaxy are compact globular clusters, rich in stars. These distant groups are outside the galaxy disc, some as much as 60,000 light-years away. The hundred or so known globular clusters move at high speeds in orbits that may take 100 million years to complete.

In one nearby globular cluster astronomers have photographed over 30,000 bright stars. Yet, adding probably faint stars, the population of each globular cluster may reach into hundreds of thousands. These compact clusters are from 60 to 300 light-years in diameter. One, in the constellation Hercules, can be observed easily during the summer.

The large globular cluster M 13 in the constellation Hercules, about 30,000 light-years away, contains some 500,000 stars.

Lastly, among our galaxy clusters are associations of very young hot stars in the thin spiral arms of the disc. Only about a hundred associations are known. The Great Nebula in the sword of Orion is one of them.

globular cluster in the constellation Virgo

globular cluster in the constellation Centaur

We might well think of our galaxy as all the universe. In it are perhaps a hundred billion stars, some new and bright, some old, small, and denser than anything else we know. This great galaxy ball, 100,000 light-years or so in diameter, is mainly empty space, yet the stars and gases in it are as heavy as a hundred billion suns. Knowing we are a tiny part of such a great system staggers our minds.

But our galaxy is by no means the end of the universe. The great optical and radio telescopes have detected millions of galaxies beyond our own. They range in distance from 200 thousand light-years to 800 billion light-years away.

Most of the millions of galaxies fall into several groups. Elliptical galaxies are most common and are somewhat smaller than

The Magellanic Clouds are the nearest star system outside our galaxy.

others. They consist almost entirely of old stars, formed when the galaxy was very young. Larger and better known are the two types of spiral galaxies. Our galaxy and many others are normal spirals with spiral arms. About a quarter of the spiral galaxies have a bar through their center.

The nearest galaxy to us is the large Magellanic Cloud, discovered by Magellan.

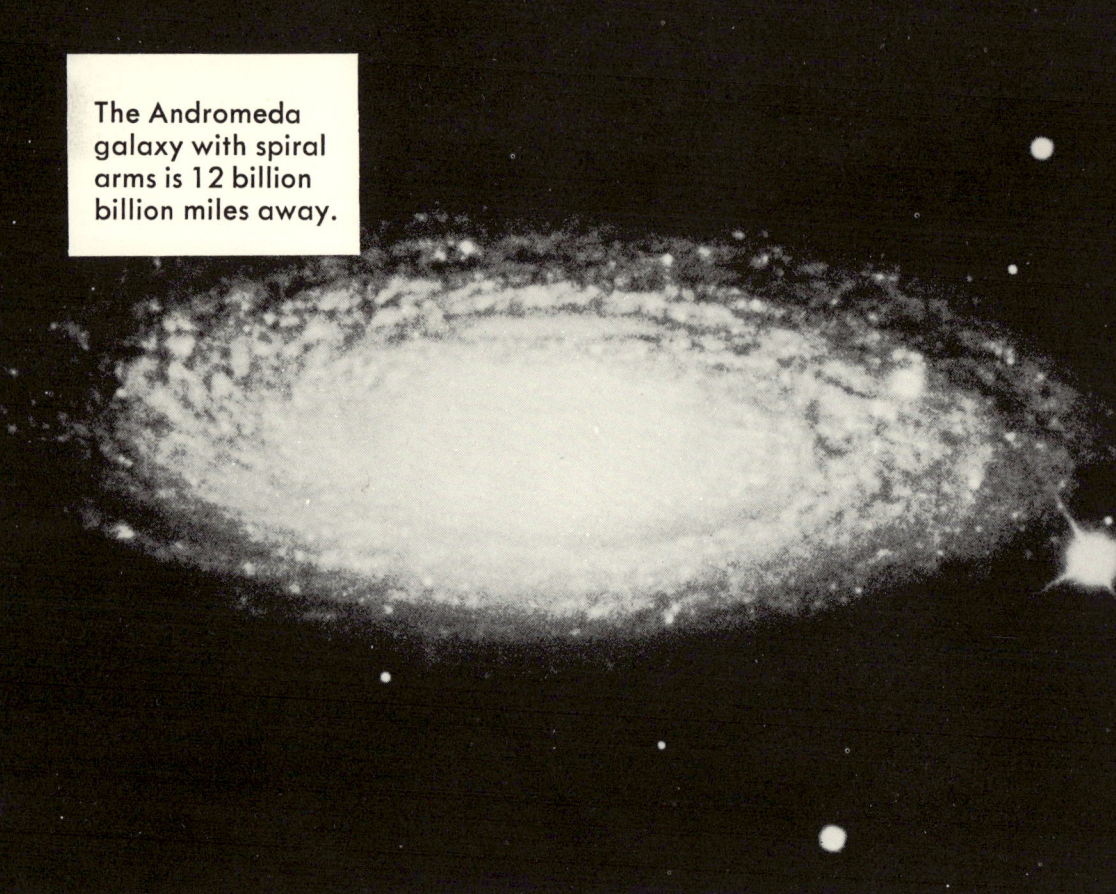

The Andromeda galaxy with spiral arms is 12 billion billion miles away.

Much smaller than our galaxy, it is about 30,000 light-years in diameter and 160,000 light-years away. This somewhat irregular galaxy rotates more slowly than ours and other spirals.

The spiral nebula of Andromeda is almost a duplicate of our own galaxy, though half

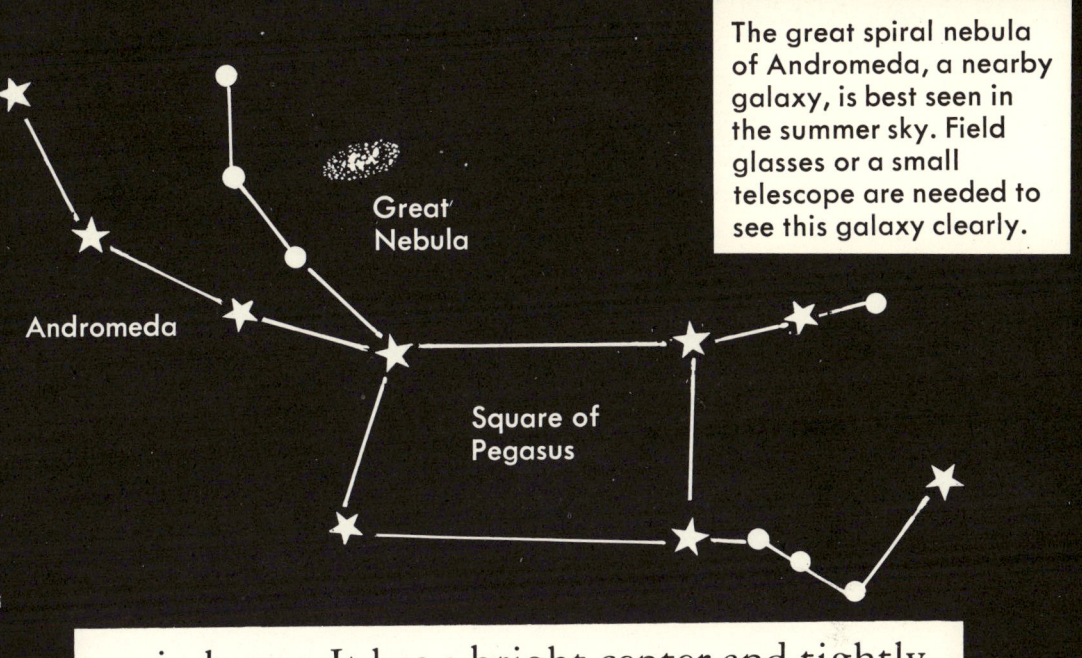

The great spiral nebula of Andromeda, a nearby galaxy, is best seen in the summer sky. Field glasses or a small telescope are needed to see this galaxy clearly.

again larger. It has a bright center and tightly spiraled arms. The Andromeda galaxy contains about the same amount of material as our galaxy and rotates in about the same way. It is over two million light-years away.

constellation Andromeda and the Spiral Nebula

The Andromeda galaxy, the Magellanic Cloud, and fifteen other galaxies including our own form a local group. This group is about three million light-years in diameter and is a typical irregular or open cluster of galaxies. Most galaxies are found in regular or rounded clusters, each with a thousand galaxies or more. Many of these star clusters can be seen with small telescopes, and some with ordinary binoculars.

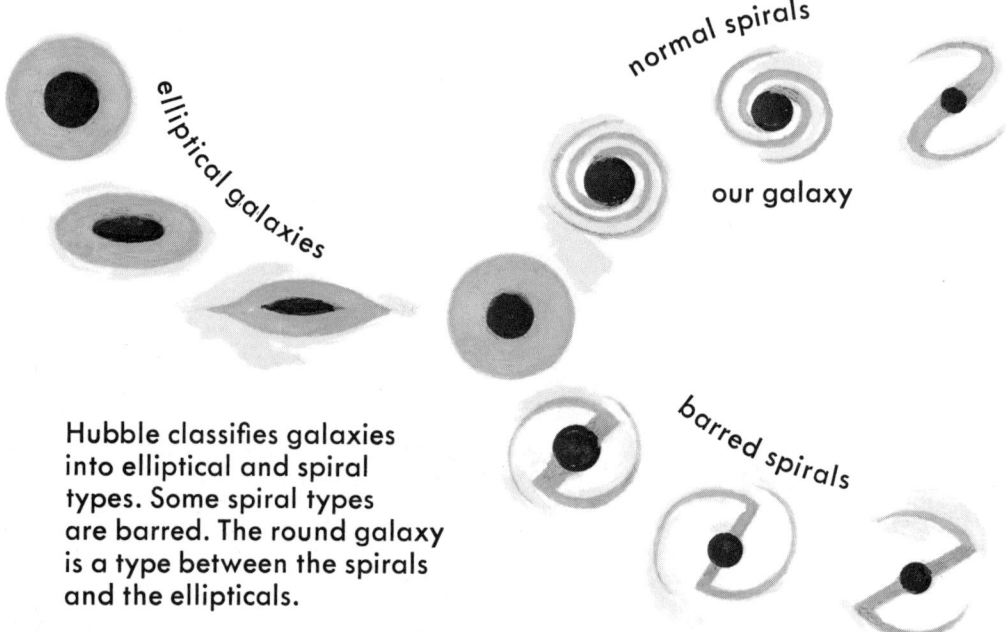

Hubble classifies galaxies into elliptical and spiral types. Some spiral types are barred. The round galaxy is a type between the spirals and the ellipticals.

New "telescopes" now reveal aspects of the universe still poorly understood. Astronomers today talk of quasars, pulsars, and black holes. They are concerned about neutron stars and invisible radio galaxies. Special photographs are made using light from hydrogen, the most common element in space. Other radiation is measured but not pictured. From it all comes a newer and better model of the universe.

The **Merry-go-round,** first radio telescope, made and used by Karl Jansky in 1931

The 250-foot radio telescope at Jodrell Bank, England, took seven years to plan and build.

These new kinds of "telescopes" detect radiation that cannot be seen, like radio waves, X rays, and gamma rays. They may come from very dim, very distant, or very special stars. The energy from these stars is often great. It may come steadily or in rapid pulses. One such star puts out a thousand times more X rays than it does visible light, an energy output equal to over 10,000 suns.

Some radio telescopes can be directed toward the source of radio waves and X rays. Larger kinds turn across the sky as the earth turns. Radio telescopes have received energy from bodies five billion light-years away. They may reach out into space for twice that distance.

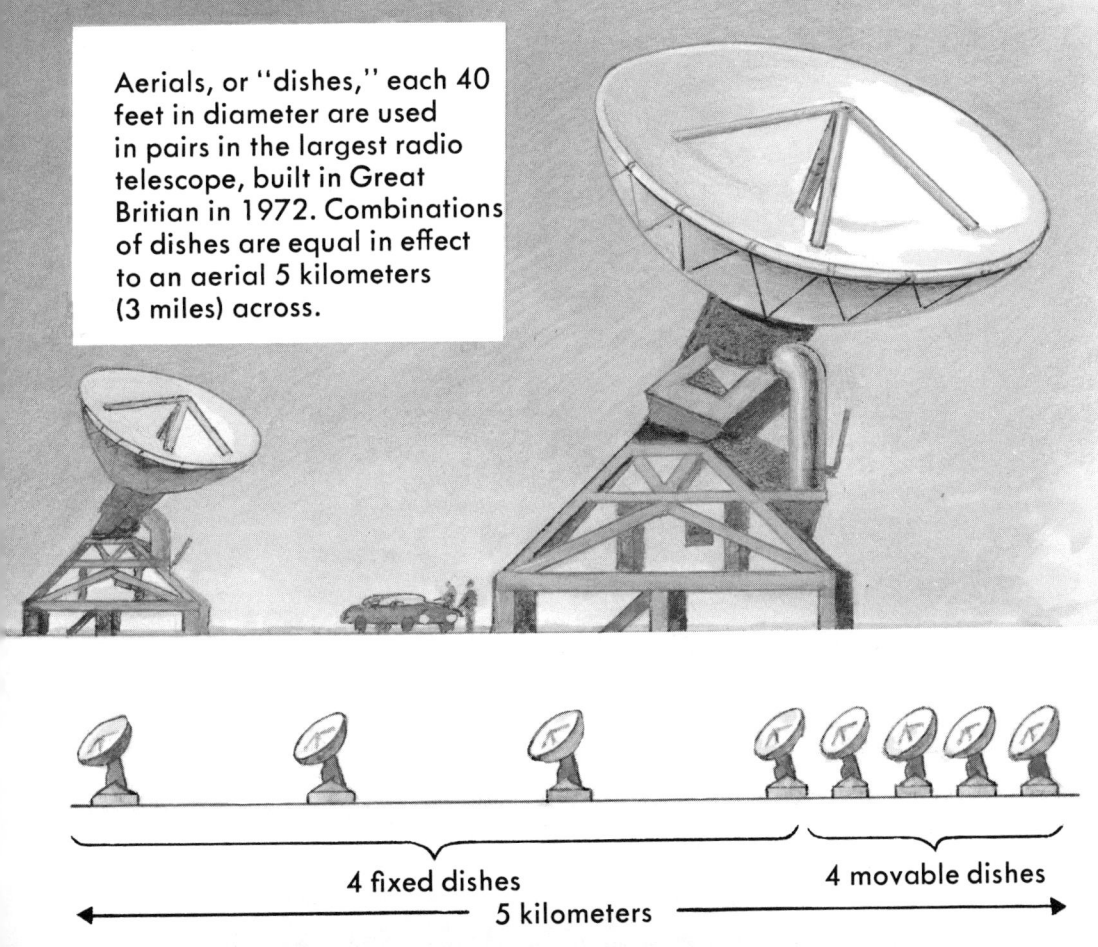

Aerials, or "dishes," each 40 feet in diameter are used in pairs in the largest radio telescope, built in Great Britian in 1972. Combinations of dishes are equal in effect to an aerial 5 kilometers (3 miles) across.

4 fixed dishes — 5 kilometers — 4 movable dishes

Telescopes that receive and record light waves on photographic film have been improved also. Some are really gigantic cameras. With one, the shift of light in a galaxy six billion light-years away has been measured. It also has photographed another galaxy eight billion light-years away.

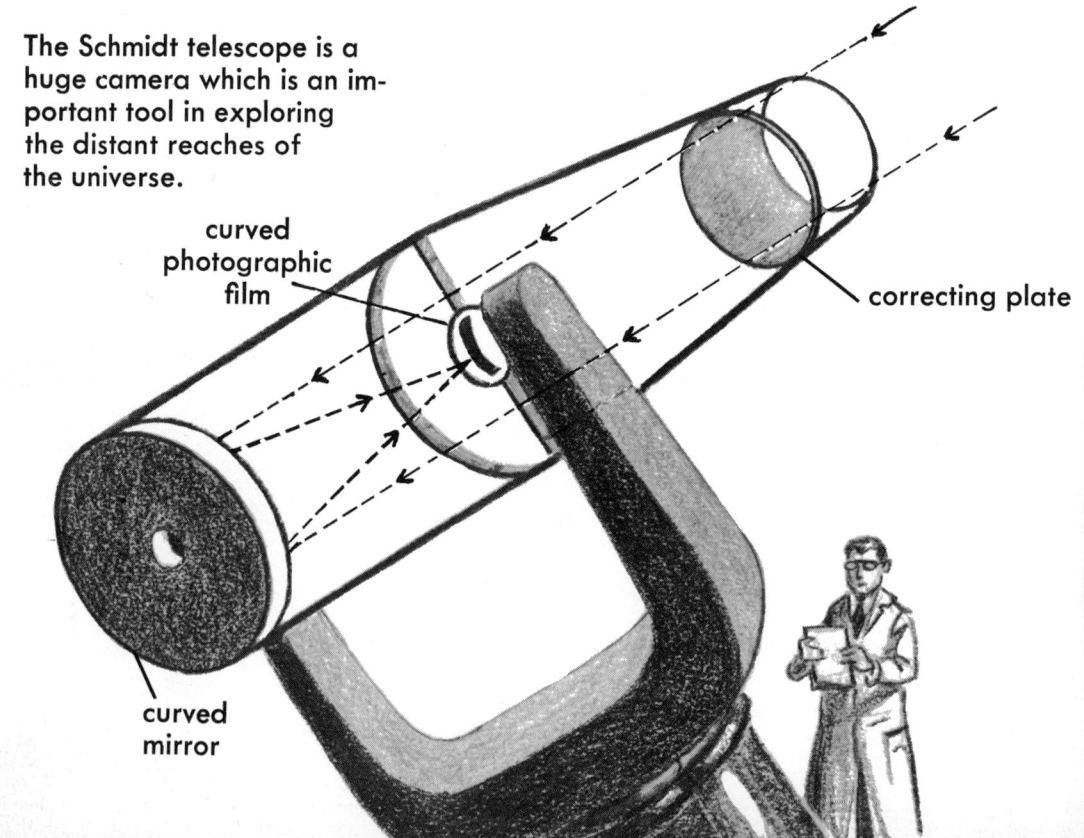

The Schmidt telescope is a huge camera which is an important tool in exploring the distant reaches of the universe.

curved photographic film

correcting plate

curved mirror

But more important than telescopes is the greatest tool of all: the mind of man. The astronomers and other scientists are the ones who observe and record and compare. From what they see they figure out the relationship between size and distance, between brightness and distance, between color and speed.

Men like Hubble, Shapley, Baade, Bok, Gamov, and others have explored the universe. Some used telescopes; some used mathematics. All used their minds to wrestle with problems that, at first, seemed impossible to solve.

OTHER GALAXIES

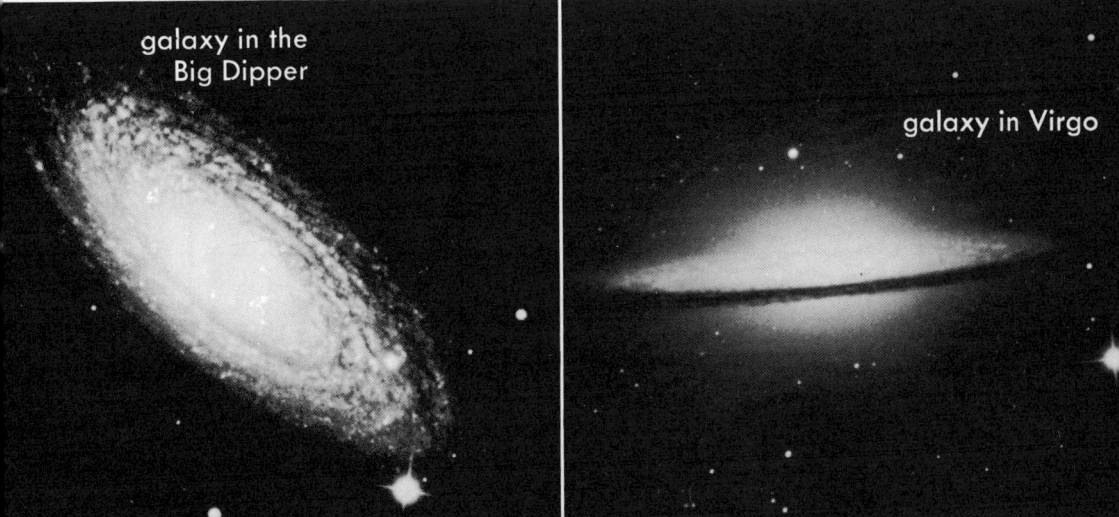

galaxy in the Big Dipper

galaxy in Virgo

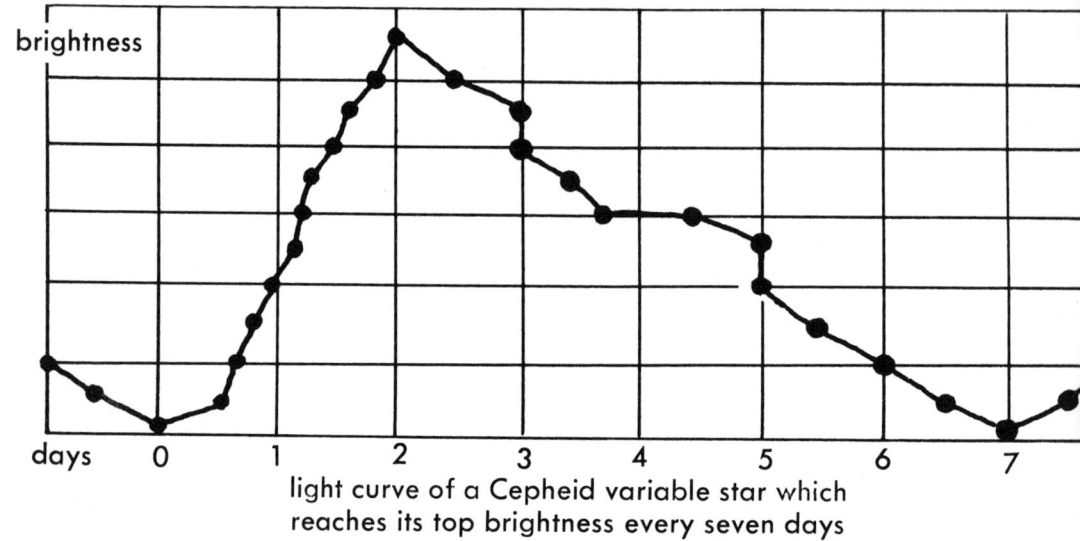

light curve of a Cepheid variable star which reaches its top brightness every seven days

In the star clouds and in other distant galaxies astronomers find variable stars. These stars change in brightness according to a regular pattern. Astronomers have learned that the changing patterns of certain variable stars (cepheids) are a clue to their distance.

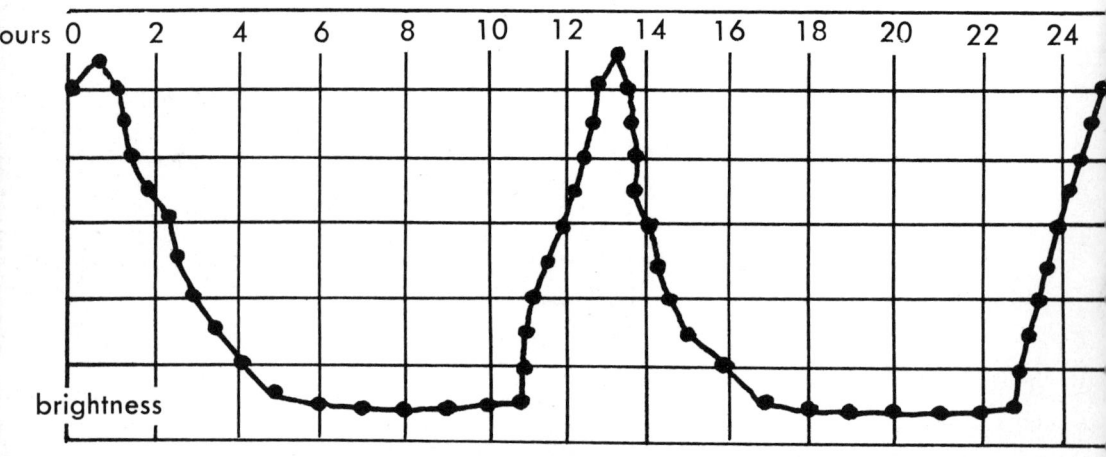

light curve of another variable star that is brightest every 13 hours

Other kinds of variable stars provide information about distance also. Some are long-period variables. Some change faster, like RR Lyrae stars, in cycles of less than one day. Most spectacular are the erupting, or explosive, stars called novas and supernovas. They increase in brightness 20 times or more before they fade away.

The Crab Nebula (M 1) is a star that exploded in July, 1054. It was as bright as the planet Jupiter at first but now is much fainter.

The Crab explosion is still going on after 900 years. Every day the Crab Nebula increases in diameter by 120 million miles. It is about 3500 light-years away.

A great handicap to astronomers is the blanket of air around our earth, and the smoke and smog in it. To avoid this, large telescopes are built on mountaintops and smaller ones are sent above the air in rockets and satellites. Scientists will put telescopes on space stations, and perhaps on the moon. This will help us understand more about how our universe began, how it is changing, whether it will ever end.

Besides pushing far into space, astronomers are also gaining information we still lack about our own small solar system. Astronauts have left important instruments on the moon. Soft landings have put down devices that send back information from Venus and Mars. Spaceships have transmitted marvelous closeup pictures of the moon, Mars, and other planets.

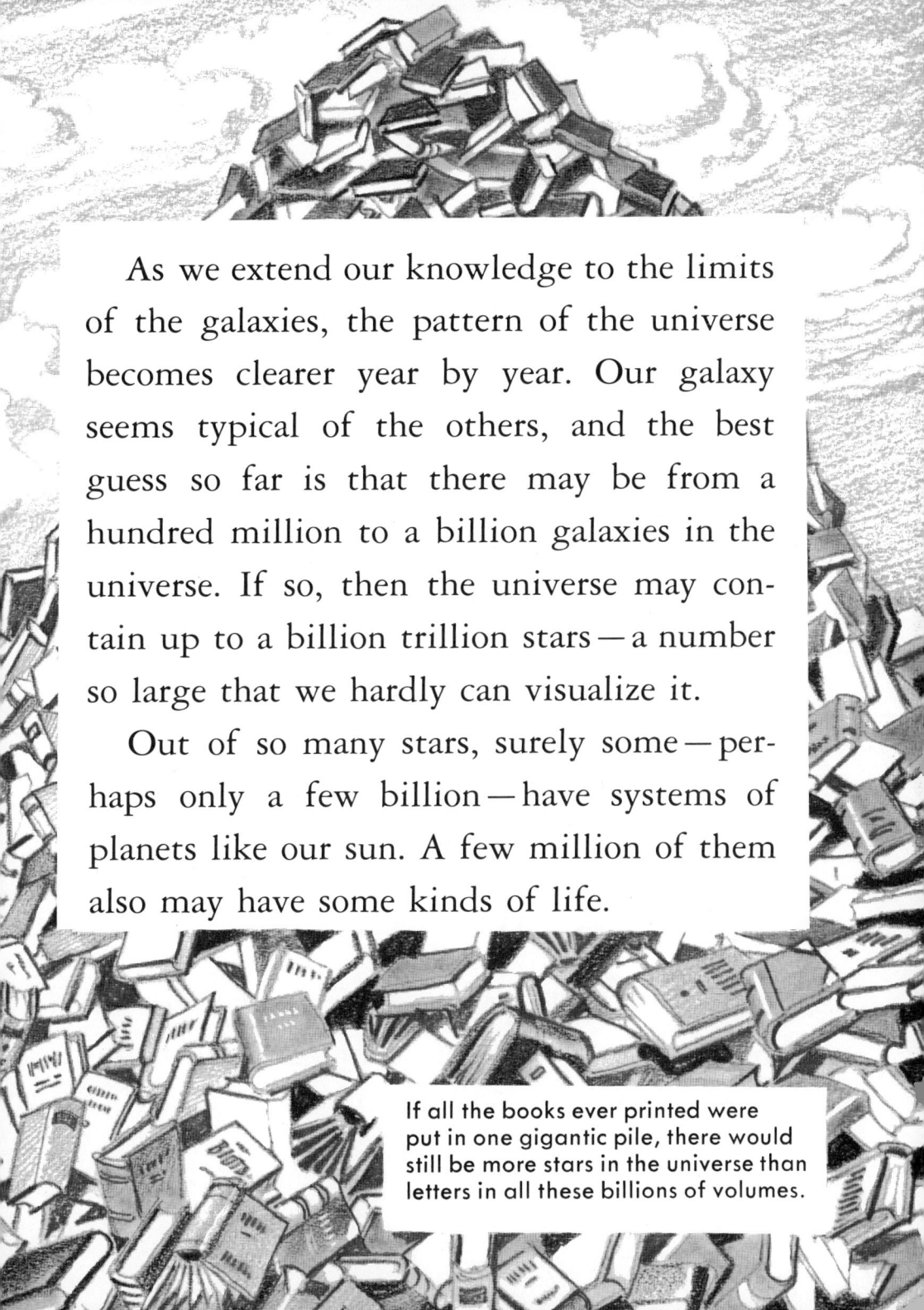

As we extend our knowledge to the limits of the galaxies, the pattern of the universe becomes clearer year by year. Our galaxy seems typical of the others, and the best guess so far is that there may be from a hundred million to a billion galaxies in the universe. If so, then the universe may contain up to a billion trillion stars — a number so large that we hardly can visualize it.

Out of so many stars, surely some — perhaps only a few billion — have systems of planets like our sun. A few million of them also may have some kinds of life.

If all the books ever printed were put in one gigantic pile, there would still be more stars in the universe than letters in all these billions of volumes.

As interesting as the chances of life around distant stars is the possibility that more distant galaxies are retreating rapidly from us and each other. They behave as if a great explosion had thrown galaxies and stars out in all directions. The farther away these bodies, the faster they seem to travel. The fastest measured so far are moving away at about 76,000 miles per second.

Faint clusters of galaxies, located by radio telescopes, indicate they are moving away from us at about half the speed of light. Nothing else in the universe travels at anything approaching the speed of these distant galaxies.

The evidence for this strange kind of movement, studied for some fifty years, is the reddening of light coming to us from far-out stars and galaxies.

THE EXPANDING UNIVERSE

Studying the light from many galaxies and clusters shows that the more distant the stars, the redder their light. The way that astronomers account for this red shift is to assume that the universe is expanding and that all parts are moving out in all directions, as shown in the chart below.

Star Cluster in	Distance in Light-Years	Rate of Moving Away	Amount of Red Shift
Virgo	50 million	750 miles per second	➡
Big Dipper	650 million	9300 miles per second	➡
Northern Crown	940 million	13,400 miles per second	➡
Herdsman	1700 million	24,400 miles per second	➡
Hydra	2600 million	38,000 miles per second	➡

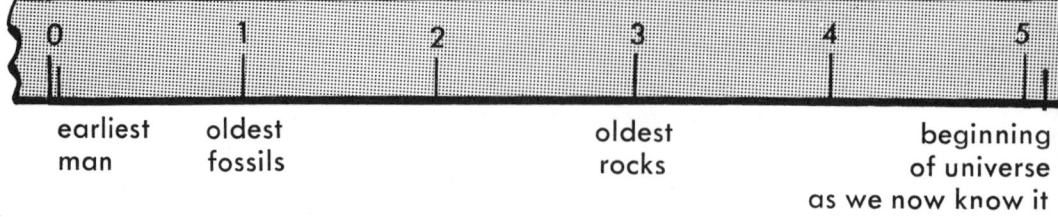

POSSIBLE TIME SCALE
in billions of years

earliest man — oldest fossils — oldest rocks — beginning of universe as we now know it

Do these observations mean that the whole universe is expanding? Some astronomers suggest that the entire universe had its beginning about five billion years ago, not much before that of the earth and the solar system. The beginning may have been a tremendous explosion. Matter that was then only atomic and subatomic in form joined together and formed the chemical elements and compounds that we now know.

Material, once formed, has been moving out from the explosion center in all directions ever since. As a small part of this process the solar system developed. If this idea is true, it was the greatest explosion ever known, and its effects are not yet over.

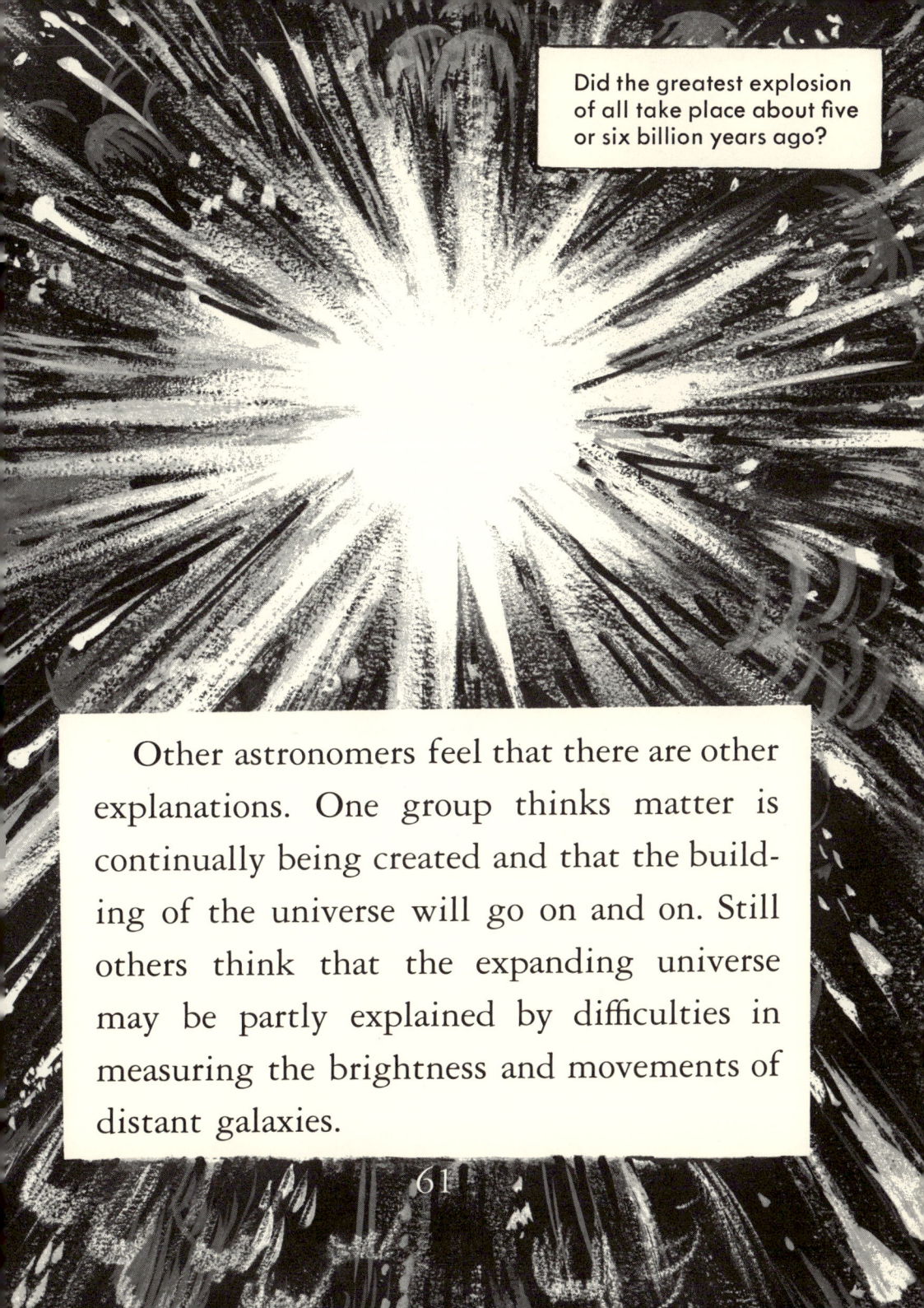

Did the greatest explosion of all take place about five or six billion years ago?

Other astronomers feel that there are other explanations. One group thinks matter is continually being created and that the building of the universe will go on and on. Still others think that the expanding universe may be partly explained by difficulties in measuring the brightness and movements of distant galaxies.

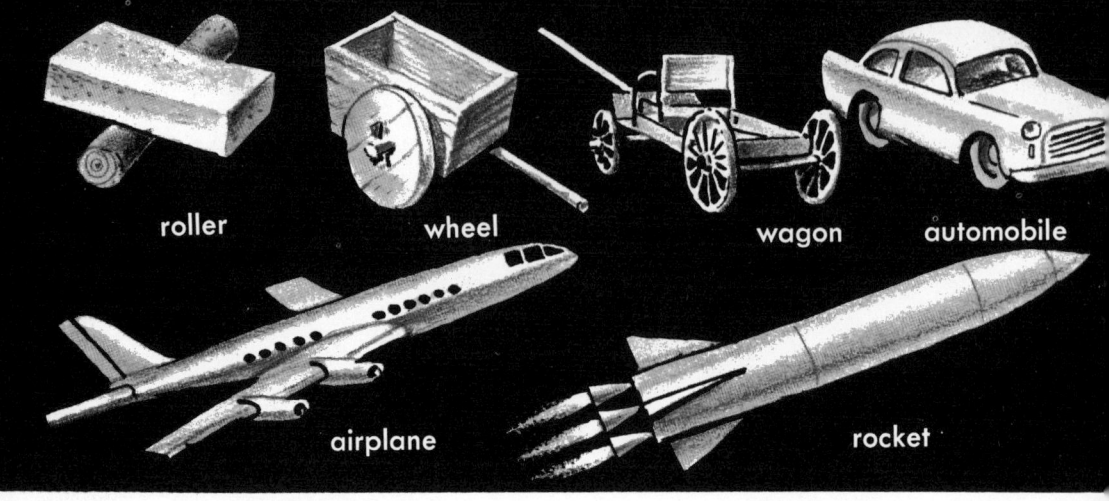

5000 years of changing ideas in transportation

Men always have wondered about the universe, but have come to understand it only during the last 5000 years. In that time great inventions and discoveries have appeared. Transportation, for example, has moved from rollers to rockets. Changes in science, industry, arts, and government have all been great. Space is no longer a barrier. The moon has become familiar. Unmanned spacecraft have landed on Venus and Mars. Now explorations much deeper into space are being planned.

5000 years of changing ideas of the universe

In 5000 years both our knowledge of the universe and our own problems have increased. Now man is able to destroy his small planet, but perhaps he will stop in time. The entire universe may be running down, very slowly, with all heat escaping into space. About this danger we can do nothing.

Our earth is no longer at the center of the universe, but the human mind still may be. It questions, discovers, explains, and keeps pushing on. Like the universe itself, it seems to have no measurable limits.

INDEX
indicates illustration

asteroids, 20
astronomical unit, 32*, 37
astronomy, history of, 11–31
Bessel, 26–27
comets, 20, 21*, 22*
constellations, 14, 23, 26, 42*, 43*, 47*
Copernicus, 18–19, 22
creation myths of the universe, 8*–11*, 9*, 10*
Earth, 12, 13*, 14*, 22*, 32*, 34*, 36*, 37*, 63*
expanding universe, 4, 58–61
galaxies, 31, 48, 57, 63*; Andromeda, 46–48, 47*; Earth, 38–42, 48*, 57; Magellanic Cloud, 45*–46, 48; other, 44, 57; type of, 44–45, 48*
Galileo, 17*, 22
Herschel, William and Carolyn, 28–29, 31
Jupiter, 14*, 17, 20, 22*, 32*, 34*
Kepler, 16, 19
light-year, 35–36, 37, 38, 42
Mars, 14*, 20, 22*, 32*, 36*, 56, 62
measurement of universe, systems of, 6*, 7, 25–27*, 26*, 32–34; units of, 32*, 33–35, 37

Mercury, 14*, 22*, 32*, 36*
meteors, 20, 21*, 22*
Milky Way, 28*–30, 31*
nebulae, 24–25*
Neptune, 20, 22*
Newton, Isaac, 19*, 22
origin of universe, theories of, 56, 60–61
parallax, 25, 37
Pleiades, 41*
Pluto, 20*, 22*
Ptolemy, 13–15, 17, 18
radiation, 49–51
Saturn, 14*, 22*, 32*
size of universe, 3, 5, 28, 63
solar system, 20–22*, 28, 63*
speed of light, 34–37
star clusters, 39, 41*–43*
stars, 20, 22–37, 57; brightness of, 24*, 54*, 55*; distance of, 6, 25–27, 32, 34, 36*, 37; life on, 57–58
sun, 12–14*, 13*, 22*, 30, 34*, 36*, 40*
telescopes, 4*, 5*, 7, 17*–19*, 18*, 24, 44, 49*–52*, 56
Uranus, 19–20, 22*
variable stars, 54–55*
Venus, 14*, 32*, 36*, 56, 62
zodiac, 14

SOME BIG NUMBERS TO REMEMBER

a million	1,000,000
a billion	1,000,000,000
a trillion	1,000,000,000,000
an astronomical unit	93,000,000 miles
a light year	5,880,000,000,000 miles
a parsec	19,200,000,000,000 miles

SALT LAKE COUNTY LIBRARY SYSTEM

RIVERTON BRANCH

```
J523.1        Zim, H. S.
Zim        (The) Universe; rev. ed.
c1973                              4.32
```

PLEASE BRING ME BACK ON TIME WITH THE DATE DUE CARD IN MY POCKET.*

*Sorry, I must charge for all lost or damaged cards.

SALT LAKE COUNTY LIBRARY SYSTEM